FANGED PITCHER PLANT

ASHLEY GISH

look!

FANGED PITCHER PLANT

CREATIVE EDUCATION • CREATIVE PAPERBACKS

Published by Creative Education
and Creative Paperbacks
P.O. Box 227, Mankato, Minnesota 56002
Creative Education and Creative Paperbacks are imprints of
The Creative Company
www.thecreativecompany.us

Design and art direction by Blue Design
Edited by Kremena Spengler

Photographs by Images by Getty Images/Beachmite Photography, 10, 20, ilbusca, 5, Maureen P Sullivan, 17, Nutkamol Komolvanich, 6–7, photo by Alan Cressler, 22, Ricardo Kuehn, 12; Shutterstock/ Pavaphon Supanantananont, 15; Unsplash/Mahosadha Ong, 8, Rapha Wilde, 14; Wikimedia Commons/Ayu (WMID), 1, David Sucianto (aka jaxsprout), 13, Günther Beck von Mannagetta und Lerchenau , 23, JeremiahsCPs, cover, 9, 18–19, 21, Philiptdotcom, 11, public domain, 14–15, Richard W Sinyem, 2, Scharmann et al., 23
Every effort has been made to contact copyright holders for material reproduced in this book. Any omissions will be rectified in subsequent printings if notice is given to the publisher.

Library of Congress Cataloging-in-Publication Datas
Names: Gish, Ashley, author.
Title: Fanged pitcher plant / Ashley Gish.
Description: Mankato, Minnesota : Creative Education and Creative Paperbacks, [2026] | Series: Look! | Includes bibliographical references and index. | Audience: Ages 6–9 | Audience: Grades 2–3 | Summary: "An exotic plant with fangs and a trap that catches bugs? Elementary-level readers will learn all about the fanged pitcher plant with this striking introduction to the unusual plant's life cycle"-- Provided by publisher.
Identifiers: LCCN 2024044287 (print) | LCCN 2024044288 (ebook) | ISBN 9798889895879 (library binding) | ISBN 9781682777534 (paperback) | ISBN 9798889896678 (ebook)
Subjects: LCSH: Nepenthes bicalcarata–Juvenile literature. | Nepenthes bicalcarata–Life cycles–Juvenile literature.
Classification: LCC QK495.N35 G57 2026 (print) | LCC QK495.N35 (ebook) |
DDC 583/.887–dc23/eng/20250109
LC record available at https://lccn.loc.gov/2024044287
LC ebook record available at https://lccn.loc.gov/2024044288

Printed in India

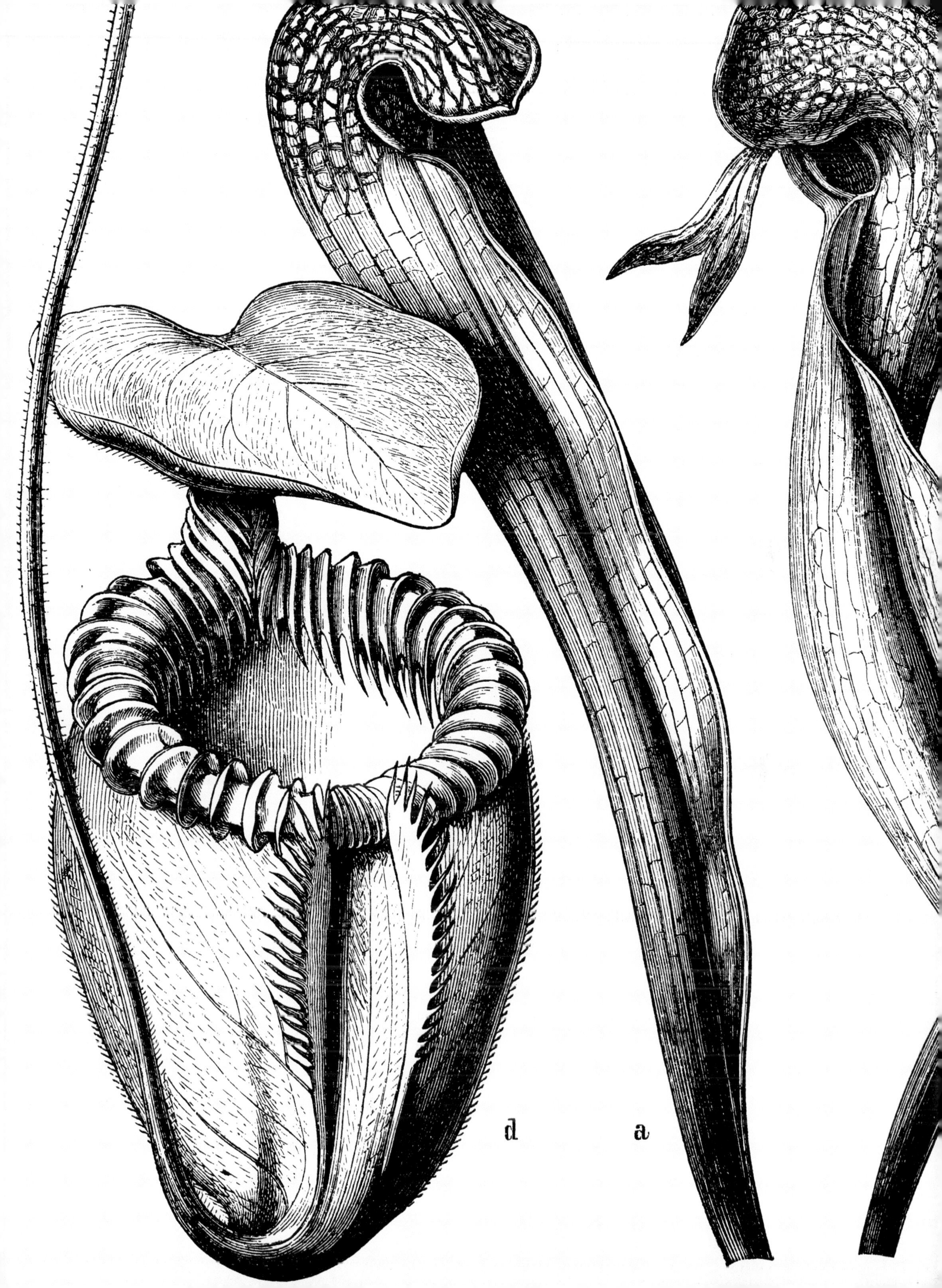
d
a

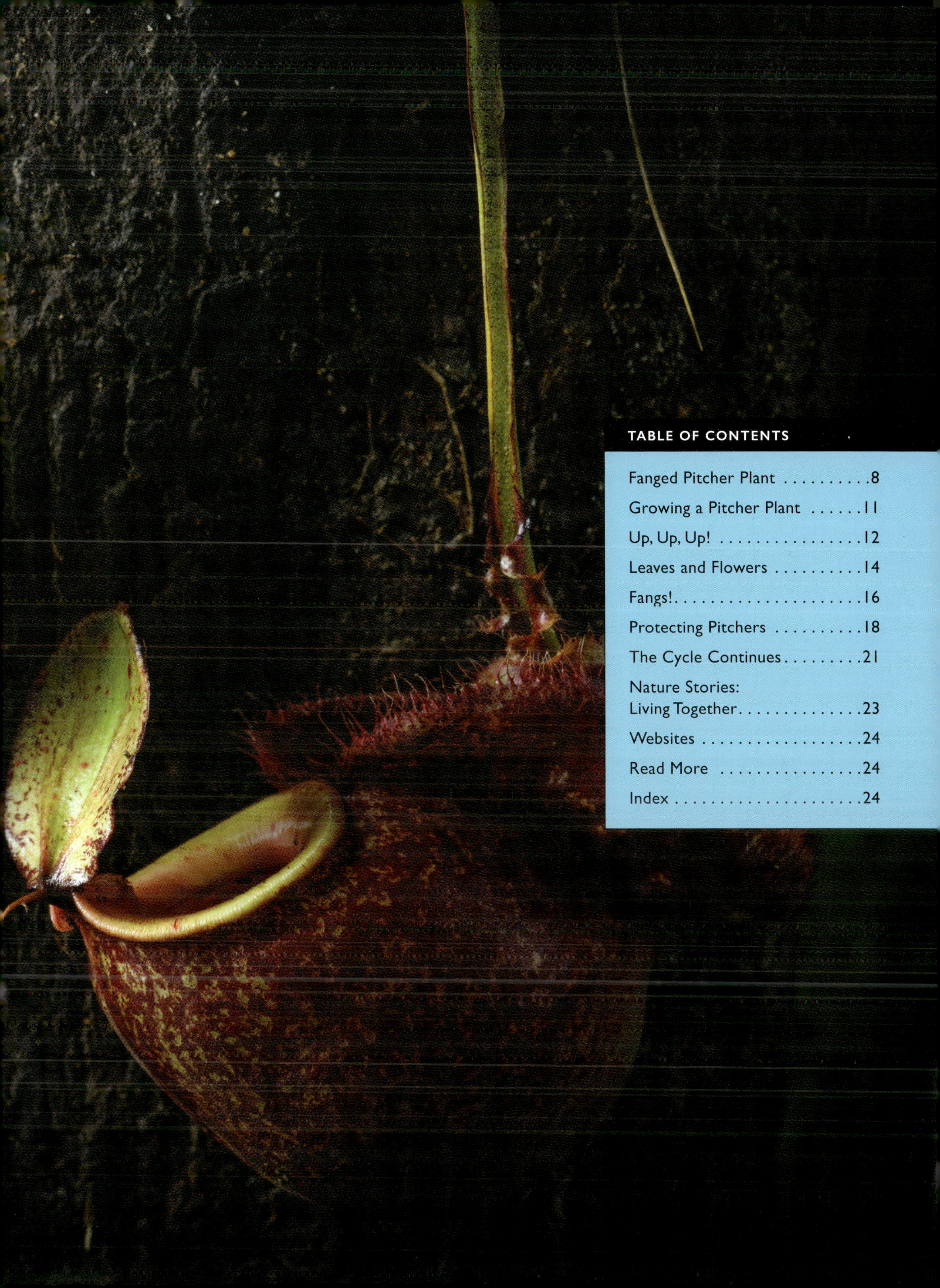

TABLE OF CONTENTS

Fanged Pitcher Plant 8
Growing a Pitcher Plant 11
Up, Up, Up! 12
Leaves and Flowers 14
Fangs! .16
Protecting Pitchers 18
The Cycle Continues 21
Nature Stories:
Living Together 23
Websites 24
Read More 24
Index 24

There are hungry mouths to feed on the island of Borneo in the Pacific Ocean. Two **fang**-like thorns grow inside the mouths. Few insects can escape them. The mouths belong to fanged pitcher plants.

Fanged pitcher plants eat bugs that fall into their pitcher-like traps. They can also break down rotten leaves and even poop! Their vines may grow taller than 49 feet (15 meters).

fang: a long, pointed tooth

Pitcher plants are named for their pitcher-shaped traps.

There are more than 100 kinds of pitcher plants.

Fanged pitcher plants grow from seeds. Wind blows the seeds away from the plant. If the dirt, water, and sunlight are just right, the seeds will grow.

Roots grow down from the seeds. One pointy leaf grows up. Soon, a **tendril** grows in the middle alongside the leaf. It takes months for more leaves to grow.

tendril: a thin, curling vine that holds onto other objects

More tendrils grow. They cling to whatever is nearby. Some begin to fatten and droop down. Pitchers grow out of the droopy tendrils.

Even young plants have pitchers. But it will take about ten years for the vine to finish growing up into the forest **canopy**.

canopy: the tops of trees in a forest

Pitchers might look like flowers, but they are actually leaves.

All fanged pitcher plants grow flowers. Some plants grow male flowers. Others grow female flowers. Flying bugs carry **pollen** from male to female flowers.

pollen: yellow powder made by plants that is used to make other plants

If pollen touches the female flowers, the flowers will make seeds. The flowers then get fat and dry. They crack open and drop their seeds.

The fangs are not really teeth. They do not chew up food. The fangs make a sweet juice called **nectar**. The nectar hangs in shining drops at the tips of the fangs. Bugs come to have a taste. Then they fall into the trap.

The edges of the trap are not slippery. If they don't fall straight down from the fangs to the trap's watery bottom, the bugs can crawl out.

nectar: sweet juice made by plants that attracts bugs and other animals

The pitcher's lid keeps rainwater out of the trap.

Fanged pitcher plants are rare in the wild. People are destroying their **habitat**. Some people dig them up to sell. This is against the law.

The good news is that fanged pitcher plants grow well in greenhouses. They can be grown from seeds. They can even be grown from a piece cut off from the parent plant!

habitat: place where a plant or animal lives

People should not take pitcher plants from the wild.

20

Life Stages
Seed
Root
Leaf
Tendril
Pitchers
Flowers

The wind carries a fanged pitcher plant seed away from its parent plant. The seed lands on good soil. It gets plenty of rainwater. The sun shines on it every day. Roots grow from the seed. A leaf grows up. Tendrils climb a nearby tree. Fanged pitchers grow from the vine. Many years later, flowers grow. If **pollinators** visit, seeds appear. The cycle starts again.

pollinator: insect that carries pollen from one flower to another

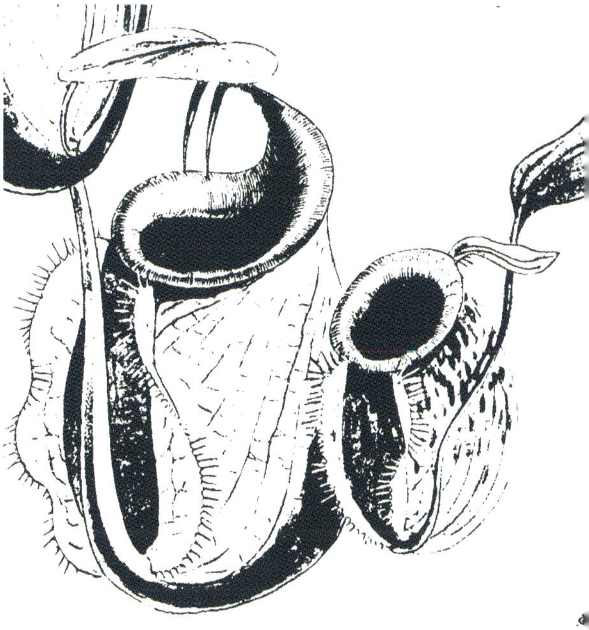

One kind of ant lives only on the fanged pitcher plant. The ants live inside the empty spaces in a plant's vines. They get food by stealing it out of the plant's trap. The plant does not eat the ants. But it won't starve. It eats the ants' poop! The plant gets energy more easily from poop than from dead bugs.

ANTS IN THE TRAP

WEBSITES

Britannica Kids: Pitcher Plant
https://kids.britannica.com/kids/article/pitcher-plant
Learn more about pitcher plants in the wild.

Carnivorous Plant Facts
https://www.edenproject.com/learn/eden-at-home/carnivorous-plants-facts
Learn more facts about carnivorous plants.

READ MORE

DK. *Plants Bite Back*. New York: DK Children, 2023.

Griffin, Mary. *Fantastic Plants: Carnivorous Pitcher Plants*. New York: Powerkids Press, 2022.

Mattern, Joanne. *Disgusting Plants*. Minnetonka, MN: Kaleidoscope, 2020.

INDEX

ant, 23
Borneo, 8
canopy, 12
fang, 8, 16
flower, 13, 14, 15, 21
habitat, 18
leaf, 11, 21
nectar, 16
pollen, 14, 15, 21
pollinator, 21
seed, 11, 15, 18, 21
tendril, 11, 12, 21
trap, 8, 9, 16, 17, 23